THE
PYTHAGOREAN THEOREM

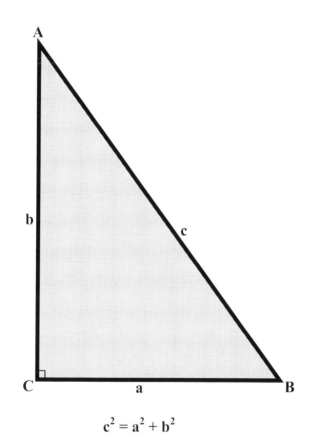

$$c^2 = a^2 + b^2$$

Merle A. Barlow

WESTBOW
PRESS®
A DIVISION OF THOMAS NELSON
& ZONDERVAN

WestBow Press books may be ordered through booksellers or by contacting:

WestBow Press
A Division of Thomas Nelson & Zondervan
1663 Liberty Drive
Bloomington, IN 47403
www.westbowpress.com
1 (866) 928-1240

ISBN: 978-1-5127-2752-4 (sc)
ISBN: 978-1-5127-2754-8 (hc)
ISBN: 978-1-5127-2753-1 (e)

Library of Congress Control Number: 2016900877

Print information available on the last page.

WestBow Press rev. date: 02/05/2016

This Book is dedicated to all my right-angled friends, and any others who may be either uncomfortable or inquisitive regarding the hypotenuse.

Preface

As the crowning achievement of the Sovereign Lord God's creations, we humans are blessed with significant characteristics. Among the blessings that separate us from His animal creations are the gifts of an eternal soul, articulate speech, and the ability to reason (i.e., the intellectual process of seeking truth or knowledge by inferring either fact or logic).

This gift of intelligence has enabled us to <u>discover</u> many exciting and wonderful concepts that the Lord has designed into His Reality.

The Pythagorean Theorem is merely one of an indefinable number of these concepts.

Clearwater, Florida
February, 2016

Merle

Prologue

My Rubric for our American Culture

Technology is good, but personal relationships (NOT social media) are superior, and books always work.

Note to Reader:

CONTENTS

CONTENTS

C O N T E N T S

1
Definition

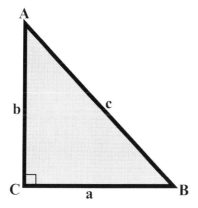

T he hypotenuse (c) of a right triangle is the side opposite the right angle. The hypotenuse is always the longest side of the right triangle.

The Pythagorean Theorem
The square of the hypotenuse is equal to the sum of the squares of the other two sides.
$$c^2 = a^2 + b^2$$

2
History

Pythagoras, a Greek Mathematician and Philosopher who lived during the fifth and fourth century before the birth of the Lord Jesus Christ, is credited with developing the theory of functions, the significance of numbers, and the Pythagorean Theorem for right triangles. However, this theorem is likely to have been known long before the time of Pythagoras. It was probably used by the ancient Egyptians in their work with the pyramids.

Actually, the Pythagorean Theorem was probably discovered multiple times in various cultures. There is evidence to indicate that this theorem was discovered by the Babylonians almost two thousand years prior to Pythagoras. There is also specific evidence that in India, a statement of the theorem and its proof existed. The theorem and its applications also were recorded in early Chinese history.

There are hundreds of proofs for the Pythagorean Theorem. Some proofs are developed by the analysis of geometric areas, some are algebraic, and some use other forms of mathematics. The interest in this mathematical relationship is perhaps greater than any other theorem in mathematics.

3
Applications

Right angles and right triangle relationships are prevalent throughout the universe and our world because of the creative design of the Sovereign Lord God, the Creator, Sustainer, and Redeemer of all Reality. Our physical world is replete with right triangle relationships developed by human skills in a broad range of creations from the simplest structures to the most complex buildings, machinery, and all types of land, sea, and air conveyances.

The Pythagorean Theorem provides a method of calculating the length of any side of a right-angled triangle if the lengths of the other two sides can be determined.

Following, is a **minimal** and **incomplete** list of occupations that utilize right triangle relationships:
Architect
Astronomer
Carpenter
Draftsman
Engineer
Interior Designer
Mechanic
Photographer
Pilot

4
Area Representation of Pythagorean Theorem

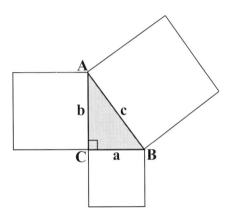

In the diagram above, $a^2 + b^2 = c^2$
$$3^2 + 4^2 = 5^2$$
The numerical values of a, b, and c validate the equation.

The sum of the <u>areas</u> of the squares on a and b equal to the <u>area</u> of the square on c also validates the equation.
$$\textbf{9 units}^2 + \textbf{16 units}^2 = \textbf{25 units}^2$$

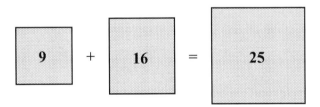

5
Example Calculations

Example 1

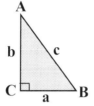

Given: a = 3; b = 4
Solve for c

$$c^2 = a^2 + b^2$$
$$= (3)^2 + (4)^2$$
$$= 9 + 16$$
$$= 25$$
$$c = \sqrt{25}$$
$$c = 5$$

Note that c is greater than a or b.
The hypotenuse will always be the longest of the three sides of a right triangle.

Example 2

If a triangle has sides of 7, 8, and 12, is the triangle a right triangle?

Solution:
$$c^2 = a^2 + b^2$$
$$(12)^2 = (7)^2 + (8)^2$$
$$144 = 49 + 64$$
$$144 \neq 113$$
Therefore, a right triangle cannot have sides of 7, 8. and 12. In other words, it is <u>not</u> a right triangle.

Example 3

Does a triangle with sides equal to $\sqrt{5}$, $\sqrt{7}$, and $\sqrt{12}$ have a right angle? In other words, is it a right triangle?

Solution:
$$\text{Does } \left(\sqrt{5}\right)^2 + \left(\sqrt{7}\right)^2 = \left(\sqrt{12}\right)^2$$
$$5 + 7 = 12$$
$$12 = 12$$
Yes, this triangle is a right triangle.

6
Pythagorean Triples

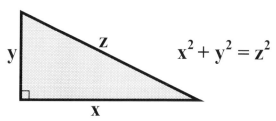

$$x^2 + y^2 = z^2$$

Given: x, y, and z are Positive Integers
Given: a and b are Positive Integers
Select a and b such that 2ab is an <u>Even</u> <u>Perfect</u> <u>Square</u>
Then, $x = a + \sqrt{2ab}$,
$\qquad y = b + \sqrt{2ab}$, and
$\qquad z = a + b + \sqrt{2ab}$

(x, y, z) is known as a Pythagorean Triple, and the numbers x, y, and z are listed in ascending sequence which indicates that "z" is the largest of the three numbers. Because it is the largest, it is <u>necessarily</u> the hypotenuse of the right triangle (x, y, z).

Example 1
Let 2ab = 100
Choose a = 5 and b = 10 so that (2)(5)(10) = 100
Therefore,
x = 5 + 10 = 15
y = 10 + 10 = 20
z = 5 + 10 + 10 = 25
The Pythagorean Triple is (15,20,25)

Example 2a
Let 2ab = 400
Choose a = 10 and b = 20 so that (2)(10)(20) = 400
Therefore,
x = 10 + 20 = 30
y = 20 + 20 = 40
z = 10 + 20 + 20 = 50
The Pythagorean Triple is (30,40,50)

Example 2b
Let 2ab = 400
Choose a = 4 and b = 50 so that (2)(4)(50) = 400
Therefore,
x = 4 + 20 = 24
y = 50 + 20 = 70
z = 4 + 50 + 20 = 74
The Pythagorean Triple is (24,70,74)

Example 3
Let 2ab = 17956 (134^2 =17956)
Choose a = 2 and b = 4489 so that (2)(2)(4489) = 17956
Therefore,
x = 2 + 134 = 136
y = 4489 + 134 = 4623
z = 2 + 4489 + 134 = 4625
The Pythagorean Triple is (136,4623,4625)

Note:
If 2ab is not an Even Perfect Square, then both a and b cannot be integers, and the values of
x, y, and z will not all be Integers.

Example:
Let 2ab = 9
Then ab = 9/2
Choose a = ½ and b = 9
Therefore,
x = ½ + 3 = 3½
y = 9 + 3 = 12
z = ½ +9 +3 = 12½
The Pythagorean Triple is (3½,12, 12½)

Examples of Pythagorean Triples

(3,4,5)
(5,12,13)
(6,8,10)
(7,24,25)
(8,15,17)
(9,12,15)
(9,40,41)
(10,24,26)
(14,48,50)
(15,20,25)
(24,70,74)
(30,40,50)
(136,4623,4625)

.
.
.

∞

A primitive Pythagorean triple is a triple in which x, y, and z are coprime (the greatest common factor of x, y, and z is 1).

,
(3,4,5), (5,12,13), (7,24,25), (8,15,17), (9,40,41), (11,60,61), (12,35,37), (13,84,85), (16,63,65), (20,21,29), (28,45,53), (33,56,65), (36,77,85), (39,80,89), (48,55,73), (65,72,97)

7
Right Triangle Relationships

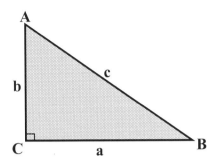

Consider the right triangle ABC with right angle C.
By the Pythagorean Theorem,

$c^2 = a^2 + b^2$, and
$c = \sqrt{a^2 + b^2}$

Question:
What is the relationship between (a + b) and c?
In other words, define this relationship mathematically.
Hint: There are exactly three possibilities, and only one is true.
Solution:
(1) Assume that $(a+b) = c$
Then $(a+b) = \sqrt{a^2 + b^2}$
Therefore, $a^2 + 2ab + b^2 = a^2 + b^2$
This is <u>false</u> because $a^2 + 2ab + b^2$ is greater by $2ab$

(2) Assume that $(a+b) < c$
Then $(a+b) < \sqrt{a^2 + b^2}$
Therefore, $(a^2 + 2ab + b^2) < a^2 + b^2$
This is <u>false</u> because $a^2 + 2ab + b^2$ is greater by $2ab$

(3) Assume $(a+b) > c$
Then $(a+b) > \sqrt{a^2 + b^2}$
Therefore, $(a^2 + 2ab + b^2) > a^2 + b^2$ by $2ab$
This is TRUE.

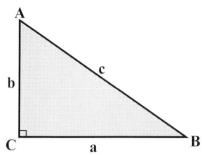

As defined on the previous page, if c denotes the length of the hypotenuse, and a and b denote the lengths of the other two sides, the Pythagorean Theorem can be expressed as:

$$a^2 + b^2 = c^2$$

If the length of both a and b are known, then c can be calculated as:

$$c = \sqrt{a^2 + b^2}$$

If the length of the hypotenuse c and either side (a or b) are known, then the length of the other side can be calculated as:

$$a = \sqrt{c^2 - b^2}$$

or

$$b = \sqrt{c^2 - a^2}$$

The Pythagorean equation relates the sides of a right triangle so that if the lengths of any two sides are known, the length of the third side can be determined.

In any right triangle, the hypotenuse is greater than either side, but less than their sum.

A triangle is determined to be right, acute, or obtuse according to the following relationships:

(1) If $a^2 + b^2 = c^2$, then the triangle is right.

(2) If $a^2 + b^2 > c^2$, then the triangle is acute.

(3) If $a^2 + b^2 < c^2$, then the triangle is obtuse.

8
Special Right Triangles

Theorem 1

In an Isosceles Right Triangle, the hypotenuse is equal to either of the two equal sides times $\sqrt{2}$.

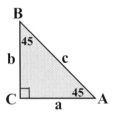

Given: Right Triangle ABC, $\angle C = 90°$, $a = b$, and
$\quad\quad\quad \angle A = \angle B = 45°$

$$c^2 = a^2 + a^2$$
$$c = \sqrt{a^2 + a^2}$$
$$c = \sqrt{2a^2}$$
$$c = a\sqrt{2}$$

The hypotenuse of an isosceles right triangle is equal to either side times $\sqrt{2}$.

For example:
Let $a = b = 7$,
Then $c = 7\sqrt{2}$

Theorem 2

In an Isosceles Right Triangle, either of the equal sides is equal to the hypotenuse divided by $\sqrt{2}$.

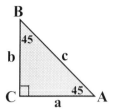

By Theorem 1 above, $c = a\sqrt{2}$.

Therefore, $a = \dfrac{c}{\sqrt{2}}$.

In an Isosceles Right Triangle, either of the two equal sides is equal to the hypotenuse divided by $\sqrt{2}$.

For example:

If $c = 7\sqrt{2}$,

Then $a = \dfrac{7\sqrt{2}}{\sqrt{2}} = 7$.

Theorem 3

In a Right Triangle, if one angle is $30°$, the hypotenuse is twice the side opposite the $30°$ angle.

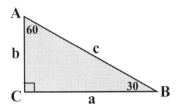

Given: Right Triangle ABC, $\angle C = 90°, \angle A = 60°, \angle B = 30°$
By Theorem 3, $c = 2b$

For example,
If $b = 8$,
Then $c = 16$.

Theorem 4

If one angle of a Right Triangle is $30°$, the side opposite the $60°$ angle is equal to $\sqrt{3}$ times the side opposite the $30°$ angle.

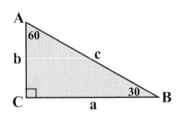

$c^2 = a^2 + b^2$

$(2b)^2 = a^2 + b^2$

$4b^2 = a^2 + b^2$

$a^2 = 3b^2$

$a = \sqrt{3b^2}$

$a = b\sqrt{3}$

For example,
If $b = 8$,
Then $a = 8\sqrt{3}$.

Theorem 5

If one angle of a Right Triangle is $30°$, the side opposite the $30°$ angle is equal to the side opposite the $60°$ angle divided by $\sqrt{3}$.

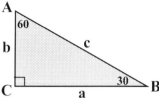

By Theorem 4, $a = b\sqrt{3}$.

Therefore, $b = \dfrac{a}{\sqrt{3}}$.

For example,

If $a = 8\sqrt{3}$,

Then $b = \dfrac{8\sqrt{3}}{\sqrt{3}} = 8$.

9
Law of Cosines

The law of cosines enables the calculation of the length of any side of <u>any</u> triangle when the length of the other two sides and the angle between them are known.

(1) $c^2 = a^2 + b^2 - 2ab\cos C$

If the angle C between the other two sides is a right angle, the law of cosines becomes the Pythagorean equation.

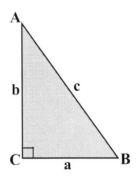

If $C = 90°$, and a and b are known, then equation (1) becomes

$c^2 = a^2 + b^2 - 2ab\cos 90°$

$c^2 = a^2 + b^2 - 2ab(0)$

$c^2 = a^2 + b^2$

$c = \sqrt{a^2 + b^2}$

10
Distance Formula

The Distance Formula in the Cartesian plane is derived from the Pythagorean Theorem. If any two points in a plane have the coordinates (x_1,y_1) and (x_2,y_2), then the distance between these two points is determined by the following formula:

$$d = \sqrt{(x_2 - x_1)^2 + (y_2 - y_1)^2}$$

On the following page, an example is illustrated, including the calculation necessary to determine the distance between the given points.

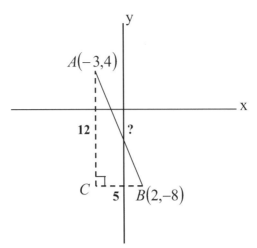

Given: $A(-3,4), B(2,-8)$

Determine: $\overline{AB}$

$$\left(\overline{AB}\right)^2 = \left(x_2 - x_1\right)^2 + \left(y_2 - y_1\right)^2$$

$$\overline{AB} = \sqrt{\left(x_2 - x_1\right)^2 + \left(y_2 - y_1\right)^2}$$

$$\overline{AB} = \sqrt{\left(2 - (-3)\right)^2 + \left(-8 - 4\right)^2}$$

$$\overline{AB} = \sqrt{\left(2 + 3\right)^2 + \left(-12\right)^2}$$

$$\overline{AB} = \sqrt{25 + 144}$$

$$\overline{AB} = \sqrt{169}$$

$$\overline{AB} = 13$$

11
Merlen Formula

The Merlen formula was developed to determine the validity of the six variables of <u>any</u> triangle. The General Formula relates all six variables of a triangle by equating an expression of the three sides with an expression of the three angles. By substituting the values of the sides and angles into the general formula, the accuracy of the variables can be verified. If there is at least one incorrect variable, note that the General Formula will not indicate which particular variables are incorrect, but only that the values of the six variables are inconsistent.

$$\frac{b^2 + c^2}{a^2} = 1 + \frac{2\sin B \sin C \cos A}{\sin^2 A} \qquad \text{[General Formula]}$$

Using the Merlen Formula, the validity of the six variables for <u>any</u> right triangle can be verified. This means that given any Pythagorean triple, and the appropriate three angles, the formula will equate the triangle's three sides with the triangle's three angles.

Let $a = \sqrt{3}, b = 1,$ and $c = 2$

For an example, I will use the triple $(1, \sqrt{3}, 2)$ as the three sides, and $30°, 60°, 90°$ as the three angles consistent with these three sides. When these six variables are substituted into the Merlen Formula, the result is equality.

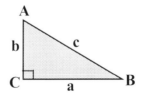

$$\frac{1+4}{3} = 1 + \frac{2\sin(30°)\sin(90°)\cos(60°)}{\sin^2(60°)}$$

$$\frac{5}{3} = 1 + \frac{2}{3}$$

$$\frac{5}{3} = \frac{5}{3}$$

12
The Pythagorean Theorem In Three Dimensions

In consideration of solid geometry concepts, the Pythagorean Theorem can be applied to three dimensions. Consider the rectangular solid illustrated below. This three-dimensional figure is also known as a rectangular parallelepiped.

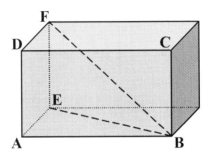

The front face of this solid is ABCD.

The left face of this solid is AEFD.

$\overline{BE}$ is the diagonal of the bottom face, and is the hypotenuse of the right triangle BAE with right $\angle$ BAE.

Applying the Pythagorean theorem to the triangle,

(1) $\left(\overline{BE}\right)^2 = \left(\overline{AB}\right)^2 + \left(\overline{AE}\right)^2$

$\overline{BF}$ is the diagonal of the solid, and is the hypotenuse of the right triangle BEF with right $\angle$ BEF.

Applying the Pythagorean theorem to this triangle,

(2) $\left(\overline{BF}\right)^2 = \left(\overline{BE}\right)^2 + \left(\overline{EF}\right)^2$

Combining equations (1) and (2), the equation for the diagonal of the solid is determined by

(3) $\left(\overline{BF}\right)^2 = \left(\overline{AB}\right)^2 + \left(\overline{AE}\right)^2 + \left(\overline{EF}\right)^2$

Taking the square root of both sides of equation (3) results in equation (4).

$$\sqrt{\left(\overline{BF}\right)^2} = \sqrt{\left(\overline{AB}\right)^2 + \left(\overline{AE}\right)^2 + \left(\overline{EF}\right)^2}$$

(4) $\overline{BF} = \sqrt{\left(\overline{AB}\right)^2 + \left(\overline{AE}\right)^2 + \left(\overline{EF}\right)^2}$

Equation (4) is a three-dimensional expression of the diagonal of the rectangular solid.

13
Right Triangle Trigonometry

The Six Trigonometric Functions

In this chapter, the trigonometric functions are presented from a right triangle perspective. Consider a right triangle, one of whose acute angles is labeled θ, as illustrated in Diagram 1. Relative to the angle θ, the three sides of the triangle are the hypotenuse, the opposite side (the side opposite the angle θ), and the adjacent side (the side adjacent to the angle θ).

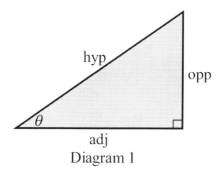

adj

Diagram 1

Using the lengths of these three sides, six ratios can be defined as the six trigonometric functions of the acute angle θ.

sine   cosecant   cosine   secant   tangent   cotangent

In the following definition, it is important to acknowledge that all angles such that $0° < \theta < 90°$, the value of each trigonometric function is positive.

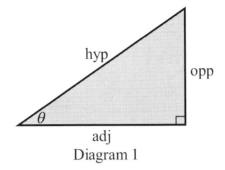

Diagram 1

Right Triangle Definitions of Trigonometric Functions

The six trigonometric functions of the angle θ are defined as follows:

$$\sin \theta = \frac{opp}{hyp} \qquad \cos \theta = \frac{adj}{hyp} \qquad \tan \theta = \frac{opp}{adj}$$

$$\csc \theta = \frac{hyp}{opp} \qquad \sec \theta = \frac{hyp}{adj} \qquad \cot \theta = \frac{adj}{opp}$$

The abbreviations opp, adj, and hyp represent the lengths of the three sides of a right triangle.

opp = the length of the side opposite θ
adj = the length of the side adjacent to θ
hyp = the length of the hypotenuse

Observe that the functions in the second row above are the reciprocals of the corresponding functions in the first row.

Use the triangle in Diagram 2 to find the values of the six trigonometric functions of θ.

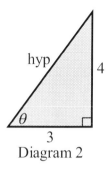

3

Diagram 2

Solution:

By the Pythagorean theorem, $(hyp)^2 = (opp)^2 + (adj)^2$, so

$$hyp = \sqrt{4^2 + 3^2} = \sqrt{25} = 5$$

Therefore, the six trigonometric functions of θ are:

$$\sin\theta = \frac{opp}{hyp} = \frac{4}{5} \qquad \cos\theta = \frac{adj}{hyp} = \frac{3}{5} \qquad \tan\theta = \frac{opp}{adj} = \frac{4}{3}$$

$$\csc\theta = \frac{hyp}{opp} = \frac{5}{4} \qquad \sec\theta = \frac{hyp}{adj} = \frac{5}{3} \qquad \cot\theta = \frac{adj}{hyp} = \frac{3}{4}$$

In Diagram 2 of the previous page, the lengths of two sides of the right triangle were given, but not the angle θ. It is more common in trigonometry to determine the trigonometric function of a given acute angle θ.

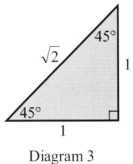

Diagram 3

Evaluating Trigonometric Functions of 45 Degrees

Find the values of $\sin 45°$, $\cos 45°$, *and* $\tan 45°$.

Solution:

Consider a right triangle having $45°$ as one of its acute angles, as illustrated in Diagram 3. The length of the adjacent side is 1. From angle relationships in triangles, the other acute angle must also be $45°$. Therefore, the triangle is isosceles, and the length of the opposite side is also 1. Using the Pythagorean Theorem, the length of the hypotenuse is $\sqrt{2}$.

$$\sin 45° = \frac{opp}{hyp} = \frac{1}{\sqrt{2}} = \frac{\sqrt{2}}{2} \qquad \cos 45° = \frac{adj}{hyp} = \frac{1}{\sqrt{2}} = \frac{\sqrt{2}}{2}$$

$$\tan 45° = \frac{opp}{adj} = \frac{1}{1} = 1$$

Evaluating Trigonometric Functions of 30 & 60 Degrees

Reference the triangle illustrated in Diagram 4 to determine the values of $\sin 60°, \cos 60°, \sin 30°, \cos 30°$.

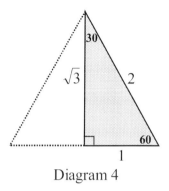

Diagram 4

Solution:

Use the triangle in Diagram 4 to verify the lengths of the given sides.

For $\theta = 60°$, $adj = 1$, $opp = \sqrt{3}$, and $hyp = 2$.
Therefore,

$$\sin 60° = \frac{opp}{hyp} = \frac{\sqrt{3}}{2} \quad \text{and} \quad \cos 60° = \frac{adj}{hyp} = \frac{1}{2}.$$

For $\theta = 30°$, $adj = \sqrt{3}$, $opp = 1$, and $hyp = 2$.
Therefore,

$$\sin 30° = \frac{opp}{hyp} = \frac{1}{2} \quad \text{and} \quad \cos 30° = \frac{adj}{hyp} = \frac{\sqrt{3}}{2}.$$

Sines, Cosines, and Tangents of Special Angles

$$\sin 45° = \frac{\sqrt{2}}{2} \qquad \cos 45° = \frac{\sqrt{2}}{2} \qquad \tan 45° = 1$$

$$\sin 30° = \frac{1}{2} \qquad \cos 30° = \frac{\sqrt{3}}{2} \qquad \tan 30° = \frac{\sqrt{3}}{3}$$

$$\sin 60° = \frac{\sqrt{3}}{2} \qquad \cos 60° = \frac{1}{2} \qquad \tan 60° = \sqrt{3}$$

Observe that $\sin 30° = \dfrac{1}{2} = \cos 60°$ This relationship is true because $30°$ *and* $60°$ are complementary angles. It can be shown from the right triangle definitions that *cofunctions of complementary angles are equal*. In other words, if θ is an acute angle, the following relationships are true:

$$\sin(90° - \theta) = \cos\theta \qquad \cos(90° - \theta) = \sin\theta$$

$$\tan(90° - \theta) = \cot\theta \qquad \cot(90° - \theta) = \tan\theta$$

$$\sec(90° - \theta) = \csc\theta \qquad \csc(90° - \theta) = \sec\theta$$

Trigonometric Identities

The study of trigonometric function relationships (Identities) is significantly important in Trigonometry.

Fundamental Trigonometric Identities

Reciprocal Identities

$$\sin\theta = \frac{1}{\csc\theta} \qquad \cos\theta = \frac{1}{\sec\theta} \qquad \tan\theta = \frac{1}{\cot\theta}$$

$$\csc\theta = \frac{1}{\sin\theta} \qquad \sec\theta = \frac{1}{\cos\theta} \qquad \cot\theta = \frac{1}{\tan\theta}$$

Quotient Identities

$$\tan\theta = \frac{\sin\theta}{\cos\theta} \qquad \cot\theta = \frac{\cos\theta}{\sin\theta}$$

Pythagorean Identities

$$\sin^2\theta + \cos^2\theta = 1 \qquad\qquad 1 + \tan^2\theta = \sec^2\theta$$

$$1 + \cot^2\theta = \csc^2\theta$$

Observe that $\sin^2\theta$ represents $(\sin\theta)^2$, $\cos^2\theta$ represents $(\cos\theta)^2$, etc.

Applying Trigonometric Identities

Let θ be an acute angle such that $\sin\theta = 0.6$. Find the values of (a) $\cos\theta$ *and* (b) $\tan\theta$ using trigonometric identities. Reference Diagram 5.

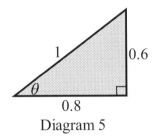

Diagram 5

Solution:

(a) To find the value of $\cos\theta$, use the Pythagorean identity

$$\sin^2\theta + \cos^2\theta = 1$$

Therefore,

$$(0.6)^2 + \cos^2\theta = 1$$

$$\cos^2\theta = 1 - (0.6)^2 = 0.64$$
$$\cos\theta = \sqrt{0.64} = 0.8$$

(b) Knowing the sine and cosine of θ, tangent θ can be determined.

$$\tan\theta = \frac{\sin\theta}{\cos\theta}$$

$$\tan\theta = \frac{0.6}{0.8}$$

$$\tan\theta = 0.75$$

Applying Trigonometric Identities

Let θ be an acute angle such that $\tan\theta = 3$. Determine the values of each trigonometric function using the trigonometric identities.

(a) $\cot\theta$ (b) $\sec\theta$

Reference Diagram 6.

Diagram 6

Solution:

(a) $\cot\theta = \dfrac{1}{\tan\theta} = \dfrac{1}{3}$

(b) $\sec^2\theta = 1 + \tan^2\theta$

$\sec^2\theta = 1 + 3^2$

$\sec^2\theta = 10$

$\sec\theta = \sqrt{10}$

Using Trigonometry to Solve a Right Triangle

Many applications of trigonometry involve a process called **solving right triangles**. In this type of application, usually one side and one acute angle of a right triangle are given, and the requirement is to determine the length of one of the other sides. Alternatively, two sides are given, and one of the acute angles must be determined.

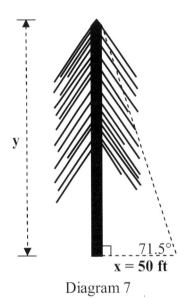

$$x = 50 \text{ ft}$$

Diagram 7

A surveyor is standing 50 feet from the base of a large Monkey tree, as illustrated in Diagram 7. The surveyor measures the angle of elevation to the top of the tree as $71.5°$. How tall is the tree? (nearest tenth of a foot)

Solution:

$$\tan 71.5° = \frac{opp}{adj} = \frac{y}{x}$$

where x = 50, and y is the height of the tree.
Therefore, the height of the tree is
$$y = x \tan 71.5°$$
$$y = 50(2.98868)$$
[Using a calculator, or a trig table for tangents]
$$y = 149.4 \, feet$$

14
A Pythagorean Trigonometric Identity

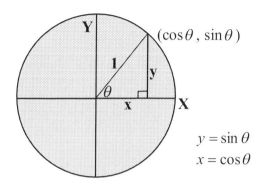

$$y = \sin\theta$$
$$x = \cos\theta$$

In a right triangle with sides x and y, and a hypotenuse that is equal to 1 (unit circle), the sine and cosine of the $\angle\theta$ between side x and the hypotenuse is:

$$\sin\theta = \frac{y}{1} = y \text{ and } \cos\theta = \frac{x}{1} = x.$$

Therefore, $y^2 + x^2 = 1^2$, or

$$\sin^2\theta + \cos^2\theta = 1$$

It is obvious that the relationship between the sine and cosine of $\angle\theta$ is an application of the Pythagorean Theorem.

15
Pythagorean Theorem Proofs

Proof 1

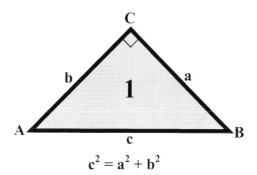

$$c^2 = a^2 + b^2$$

15.1
Algebraic Analysis of Areas I

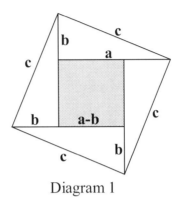

Diagram 1

Reference Diagram 1.

The **plan** is to determine the area of several polygons in two different ways, and then equate the two areas to verify the Pythagorean relationship.

Given:
A configuration of four right triangles and two squares with these specific details:
Four congruent right triangles with sides a, b, and c.
A shaded square whose side is (a-b),
A large square whose side is c.

Prove:
$$c^2 = a^2 + b^2$$

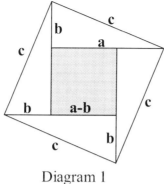

Diagram 1

Reference Diagram 1.

(1) The area of the large square is c^2.

The area of the large square consisting of the four congruent right triangles, and the small shaded square is:

$$4\left(\frac{ab}{2}\right) + (a-b)^2$$

(2) $2ab + a^2 - 2ab + b^2$

Equate areas (1) and (2):
$$c^2 = 2ab + a^2 - 2ab + b^2$$
$$c^2 = a^2 + b^2$$

35

Proof 2

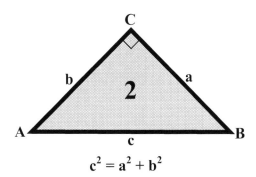

$$c^2 = a^2 + b^2$$

15.2
Algebraic Analysis of Areas II

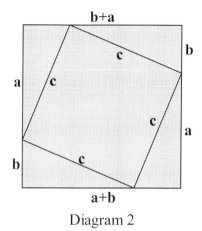

Diagram 2

Reference Diagram 2.

The **plan** is to subtract the area of several polygons from the area of a larger polygon to validate the Pythagorean relationship.

Given:
A configuration of four right triangles and two squares with these specific details:
Four congruent right triangles with sides a, b, and c.
A small square whose side is c.
A large square whose side is a+b.

Prove:
$c^2 = a^2 + b^2$

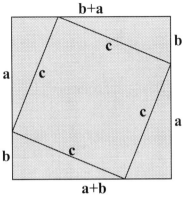

Diagram 2

Reference Diagram 2.

(1) The area of the small square with side c is c^2.

(2) The area of the large square with side (a+b) is:
$$(a+b)^2 = a^2 + 2ab + b^2$$

(3) The area of the four congruent right triangles with sides a, b, and c is:

$$4\left(\frac{ab}{2}\right)$$

Area (1) = Area (2) – Area (3)

$$c^2 = \left(a^2 + 2ab + b^2\right) - 4\left(\frac{ab}{2}\right)$$
$$c^2 = \left(a^2 + 2ab + b^2\right) - 2ab$$
$$c^2 = a^2 + b^2$$

Proof 3

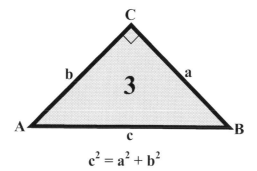

$$c^2 = a^2 + b^2$$

15.3
Algebraic Analysis of Areas III

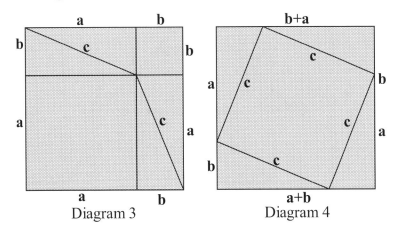

Diagram 3   Diagram 4

Reference Diagrams 3 and 4.

Proof 15.3 is similar to 15.1 and 15.2. The difference is that the composition and configuration of the polygons used in 15.3 is different from the previous proofs.

The **plan** for the solution of 15.3 is to equate two areas that will simplify to prove the Pythagorean Theorem.

Given:
In Diagram 3, there are four congruent right triangles each with sides a, b, and c
There is a small square with side b.
There is a large square with side a.
The largest square with side (a+b) is comprised of these six polygons.

In Diagram 4, there are four congruent right triangles each with sides a, b, and c.
The smallest square has a side of c.
The largest square has a side of (a+b).

Both of the largest squares of Diagram 3 and Diagram 4 have an area of $(a+b)^2$.

Prove:
$c^2 = a^2 + b^2$

40

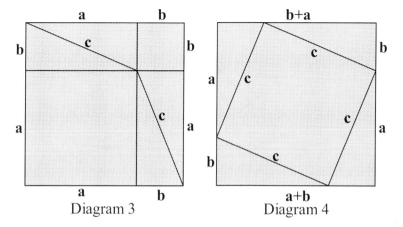

Diagram 3   Diagram 4

Reference Diagrams 3 and 4.

The total area of the six polygons of Diagram 3 is;

(1) $4\left(\dfrac{ab}{2}\right) + b^2 + a^2$

The total area of the five polygons of Diagram 4 is:

(2) $4\left(\dfrac{ab}{2}\right) + c^2$

Equating areas (1) and (2) result in:

$2ab + b^2 + a^2 = 2ab + c^2$, or

$c^2 = a^2 + b^2$

Proof 4

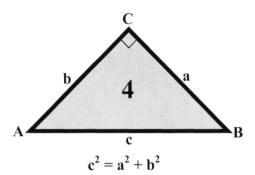

$$c^2 = a^2 + b^2$$

15.4
Trapezoid & Triangles

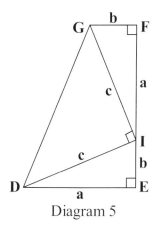

Diagram 5

Reference Diagram 5.

The **plan** for the solution of 15.4 is:
(A) Determine the area of the trapezoid by using the trapezoid area formula.
(B) Determine the sum of the areas of the three right triangles that comprise the trapezoid.
(C) Equate the two equal areas for the solution.

Given:
A trapezoid DEFG with bases a and b, and height (a+b).
Construct three right triangles in the trapezoid as illustrated. Two of these triangles will be congruent with sides a, b, and c. The third triangle will be isosceles with c for the base and c for the altitude.
Because DI (c) equals GI (c), the vertex of angle DIG is on a circle whose diameter is the side DG of the trapezoid. This means that angle DIG intercepts the diameter of a circle, and is therefore a right angle.

Prove:
$c^2 = a^2 + b^2$

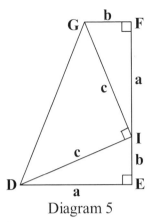

Diagram 5

Reference Diagram 5.

The area of trapezoid DEFG:

$$A_{DEFG} = \left(\frac{h}{2}\right)(DE + FG)$$

$$A_{DEFG} = \left(\frac{a+b}{2}\right)(a+b)$$

(1) $A_{DEFG} = \dfrac{a^2 + 2ab + b^2}{2}$

The total area of the three right triangles:
Area of DIE + Area DIG + Area FIG

$$A_{DIE} = \frac{ab}{2}$$

$$A_{DIG} = \frac{c^2}{2}$$

$$A_{FIG} = \frac{ab}{2}$$

(2) Total Area $= \dfrac{ab}{2} + \dfrac{c^2}{2} + \dfrac{ab}{2}$

Because Area (1) = Area (2):

$$\frac{a^2 + 2ab + b^2}{2} = \frac{ab}{2} + \frac{c^2}{2} + \frac{ab}{2}$$

$$a^2 + 2ab + b^2 = ab + c^2 + ab$$

$$a^2 + 2ab + b^2 = c^2 + 2ab$$

$$c^2 = a^2 + b^2$$

Proof 5

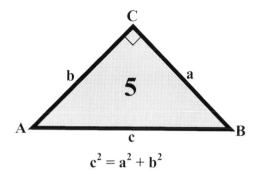

$$c^2 = a^2 + b^2$$

15.5
Similar Triangles

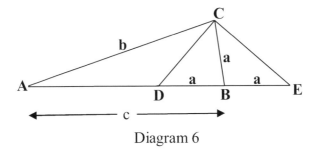

Diagram 6

Reference Diagram 6.

The **plan** for the solution of 15.5 is to prove that two triangles are similar in order to establish a proportion that will verify the Pythagorean relationship.

Given:
A triangle ABC with $\angle ACB = 90°$.
AB = c, BC = a, AC = b.
Extend AB to E to define points D and E so that
BD = BE = a.
By this definition, C lies on a circle width center B, and radius a.
$\angle DCE$ intercepts diameter DE.
Therefore, $\angle DCE = 90°$.

Prove:
$c^2 = a^2 + b^2$

46

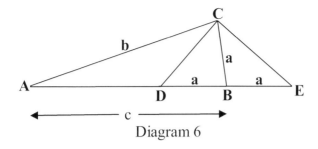
Diagram 6

Reference Diagram 6.

$\angle ACD = \angle BCE$ because both of these angles are complementary to the same angle.
Since triangle BCE is isosceles, $\angle CEB = \angle BCE$.
Triangles DAC and EAC share $\angle DAC$.
$\angle ACD = \angle AEC$ because they are both equal to the same angle.
Therefore, triangle DAC and EAC are similar, because the three corresponding angles of the two triangles are equal.
From this similarity,

$$\frac{AC}{AE} = \frac{AD}{AC}, \text{ or}$$

$$\frac{b}{c+a} = \frac{c-a}{b}$$

$$c^2 - a^2 = b^2$$

$$c^2 = a^2 + b^2$$

Proof 6

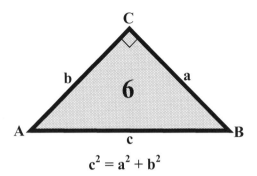

$$c^2 = a^2 + b^2$$

15.6
Intersecting Chords

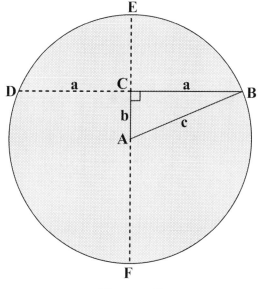

Diagram 7

Reference Diagram 7

The **plan** for the solution of 15.6 is to use the relationship of two chords that intersect within a circle.

Given:
Triangle ABC is a right triangle with $\angle ACB = 90°$.
The hypotenuse AB is used as a radius to construct the circle whose center is A.
Extend BC to a point D on the circle.
Extend AC to points E and F on the circle.

Prove:
$c^2 = a^2 + b^2$

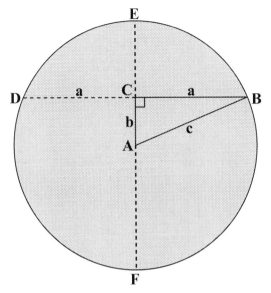

Diagram 7

Reference Diagram 7

The diameter EF is perpendicular to chord BD.
If a line through the center of a circle is perpendicular to a chord, it bisects the chord.
AF = c because like AB, AF is a radius of the circle.
Likewise, AE = c.
DC = a because DC is half of the bisected chord BD.
Chords BD and EF intersect within the circle whose center is A.
If two chords intersect within a circle, the product of the segments of one is equal to the product of the segments of the other.
Therefore,

$$(EC)(CF) = (DC)(CB)$$
$$EC = (c - b), \text{ and } CF = (c + b)$$
$$DC = CB = a$$

Hence,

$$(c - b)(c + b) = (a)(a)$$
$$c^2 - b^2 = a^2$$
$$c^2 = a^2 + b^2$$

Proof 7

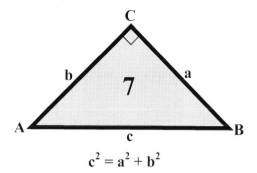

$$c^2 = a^2 + b^2$$

15.7
Orthogonal Right Triangles

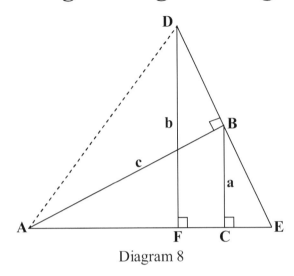

Diagram 8

Reference Diagram 8

The **plan** for 15.7 is to evaluate the area of triangle ADE in two different ways, and then equate the two equal areas to verify the Pythagorean relationship.

Given:
Triangles ABC and DEF are two congruent right triangles.
B is on DE, and F and C are on AE.
Construct AD to form triangle ADE.
BC = EF = a
AC = DF = b
AB = DE = c
$\angle ABE = \angle DFE = 90°$.

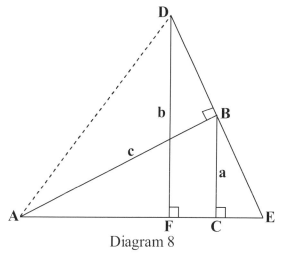

Diagram 8

Reference Diagram 8.

(1) Area of triangle ADE = $\dfrac{(AB)(DE)}{2} = \dfrac{(c)(c)}{2} = \dfrac{c^2}{2}$

(2) Area of triangle ADE = $\dfrac{(DF)(AE)}{2} = \dfrac{(b)(AE)}{2}$

$$AE = AC + CE = b + CE$$

CE can be determined by evaluating two similar triangles.
Triangles BCE and DFE are similar because each has three angles equal to the three angles of the other.
Because of the similarity,

$$\dfrac{DF}{BC} = \dfrac{FE}{CE}$$

$$\dfrac{b}{a} = \dfrac{a}{CE}$$

$$(b)(CE) = (a)(a)$$

$$CE = \dfrac{a^2}{b}$$

Therefore,

$$AE = b + \dfrac{a^2}{b}$$

Equating (1) and (2) results in

$$\dfrac{c^2}{2} = \dfrac{b\left(b + \dfrac{a^2}{b}\right)}{2}, \text{ and } c^2 = b\left(b + \dfrac{a^2}{b}\right), \text{ and}$$

$$c^2 = a^2 + b^2$$

Proof 8

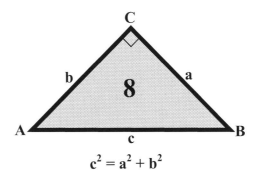

$$c^2 = a^2 + b^2$$

15.8
Hypotenuse
&
Mean Proportionals

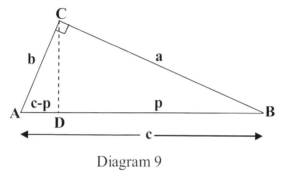

Diagram 9

Reference Diagram 9.

The **plan** is to determine the value of a^2 and b^2, and add the results.

Given:
Right triangle ABC with right $\angle C$.
Construct altitude CD from $\angle C$ to AB,
so that CD is perpendicular to AB.
DB = p, AD = (c-p).

Prove:
$c^2 = a^2 + b^2$

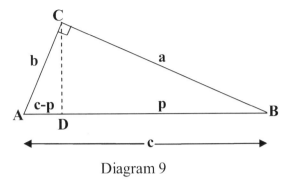

Diagram 9

Reference Diagram 9.

When the altitude upon the hypotenuse of a right triangle is constructed, either side of the triangle is the mean proportional between the hypotenuse and the segment of the hypotenuse adjacent to the side.

This is true because triangle ADC is similar to triangle ABC, and triangle BDC is similar to triangle ABC.

Hence,

$$\frac{c}{a} = \frac{a}{p}, \text{ and } \frac{c}{b} = \frac{b}{(c-p)}$$

Therefore,

(1) $a^2 = cp$, and

(2) $b^2 = c^2 - cp$

Adding (1) and (2):

$$a^2 + b^2 = cp + \left(c^2 - cp\right)$$

$$c^2 = a^2 + b^2$$

Proof 9

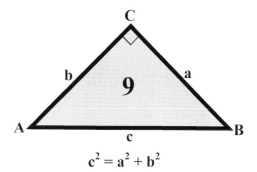

$$c^2 = a^2 + b^2$$

15.9
Congruent & Similar Triangles

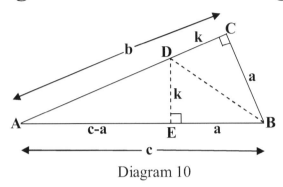

Diagram 10

Reference Diagram 10.

The **plan** for 15.9 is to use two congruent triangles and two similar triangles to establish two proportions. Adding the two proportion equations will verify the Pythagorean relationship.

Given:
Triangle ABC is a right triangle with $\angle C = 90°$.
Construct BE = BC.
At E, construct DE perpendicular to AB.
Draw BD.

Prove:
$$c^2 = a^2 + b^2$$

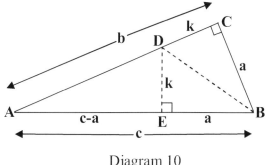

Diagram 10

Reference Diagram 10.

Triangle BDC is congruent to triangle BDE because if two right triangles have the hypotenuse and another side of one equal to the hypotenuse and a side of the other, they are congruent.

Triangles ABC and ADE are similar.

Therefore,

$$\frac{BC}{DE} = \frac{CA}{AE} \text{ and } \frac{CA}{AE} = \frac{AB}{AD}$$

Hence,

$$\frac{a}{k} = \frac{b}{c-a} \text{ and } \frac{b}{c-a} = \frac{c}{b-k}$$

From these two equations,

$$bk = ac - a^2 \text{, and}$$

$$b^2 - bk = c^2 - ac$$

Adding these equations,

$$b^2 - bk + bk = c^2 - ac + ac - a^2$$

$$b^2 = c^2 - a^2 \text{, or}$$

$$c^2 = a^2 + b^2$$

Proof 10

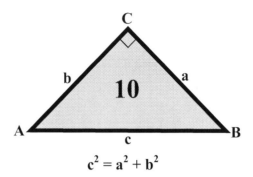

$$c^2 = a^2 + b^2$$

15.10
Secant & Tangent

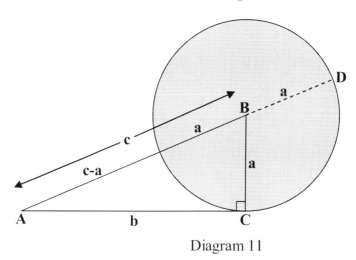

Diagram 11

Reference Diagram 11.

The **plan** is to establish a proportion involving a secant and a tangent that will verify the Pythagorean relationship.

Given:
Triangle ABC is a right triangle with $\angle C = 90°$.
BC is the radius of the circle.
Tangent AC and Secant ABD meet at an external point A.

Prove:
$c^2 = a^2 + b^2$

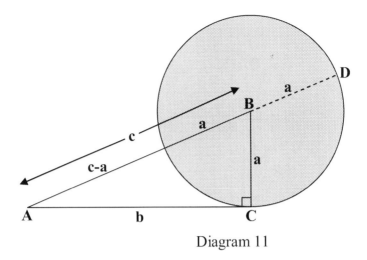

Diagram 11

Reference Diagram 11.

AC is tangent to the circle at C.
ABD is a secant whose external segment is $(c - a)$.

If a tangent and a secant intersect in the exterior of a circle, the square of the tangent is equal to the product of the secant and its external segment.

This means that the tangent is the mean proportional between the secant and its external segment.
Therefore,

$$\frac{c - a}{b} = \frac{b}{c + a}$$

$c^2 - a^2 = b^2$, or

$c^2 = a^2 + b^2$

Proof 11

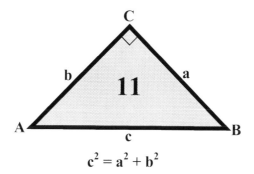

$$c^2 = a^2 + b^2$$

15.11
Area Classic

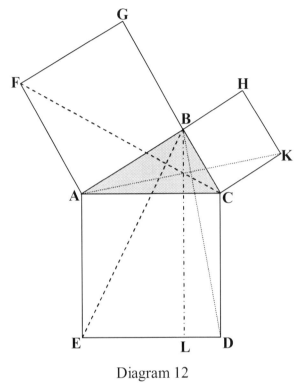

Diagram 12

Reference Diagram 12.

Following is a solution **plan** for 15.11:

AD, the largest square, is divided into a left and right rectangle by BL. A triangle is constructed that has half the area of the left rectangle. Another triangle is constructed that has half the area of square BF. These two triangles are congruent, verifying that square BF has the same area as the left rectangle.

This reasoning is repeated in a similar fashion for the right rectangle, and the smallest square, BK.

Adding the two rectangles that comprise the square on the hypotenuse, the area on the hypotenuse is determined to be the same as the sum of the areas on the other two sides.

Given:

Right triangle ABC with $\angle B = 90°$.

AD is the largest square on the hypotenuse AC.

BF and BK are the squares on sides AB and BC respectively.

Construct BL parallel to AE.

Draw lines AK and BD.

Prove:

Square AD = square BF + square BK

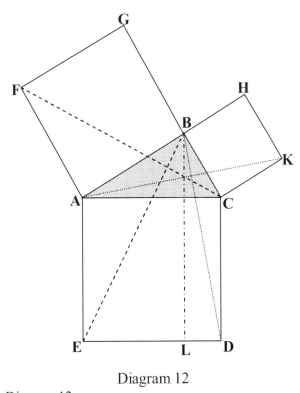

Diagram 12

Reference Diagram 12.

Triangle FAC is congruent to triangle BAE.
Triangle BAE and rectangle AL have the same base AE.
Triangle BAE and rectangle AL also have equal altitudes.
(The altitudes lie between the parallel lines AE and BL).
Therefore, rectangle AL is twice the area of triangle BAE because both have the same base and equal altitudes.
Square BF is twice the area of triangle FAC because both have the same base and equal altitudes.
Square BF is equal to the area of rectangle AL because both BF and AL are twice the area of congruent triangles FAC and BAE.

Triangle ACK is congruent to triangle DCB.
Triangle DCB and rectangle LC have the same base CD.
Triangle DCB and rectangle LC also have equal altitudes.
(The altitudes lie between the parallel lines CD and BL).
Therefore, rectangle LC is twice the area of triangle DCB because both have the same base and equal altitudes.
Square BK is twice the area of triangle ACK because both have the same base and equal altitudes.
Square BK is equal to the area of rectangle LC because both BK and LC are twice the area of congruent triangles ACK and DCB.

In summary,

Square BF = Rectangle AL
Square BK = Rectangle LC

Because Square AD = Rectangle AL + Rectangle LC,
Square AD = Square BF + Square BK

Proof 12

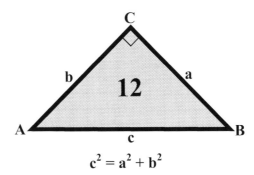

$$c^2 = a^2 + b^2$$

15.12

Square Around the Circle

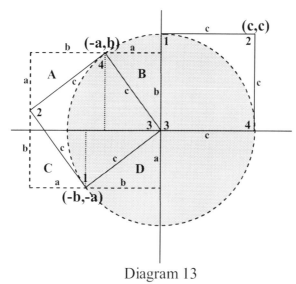

Diagram 13

Reference Diagram 13.

The **plan** is to rotate a given square with side c in quadrant I of a unit circle counterclockwise about the origin of the circle so that two of the square's vertices are now located on the circle at coordinates $(-a, b)$ and $(-b, -a)$. After the rotation, the original lower left vertex of the square is now the lower right vertex of the rotated square in its new location in quadrants II and III.

Given:
A square with side c in quadrant I of a unit circle, and c is the radius of the circle.
After the rotation,
Vertex 1 of the original square is on the circle at coordinates $(-b, -a)$.
Vertex 4 of the original square is on the circle at coordinates $(-a, b)$.
Vertex 3, the lower left vertex of the original square at the circle origin, is rotated so that it is now the lower right vertex at the origin.

Prove:
$c^2 = a^2 + b^2$

69

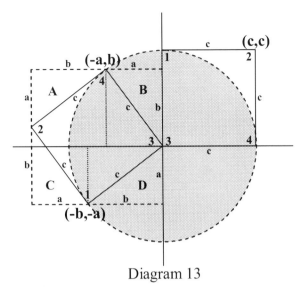

Diagram 13

Reference Diagram 13

Construct horizontal lines parallel to the x-axis through $(-a, b)$ and $(-b, -a)$ that will meet a vertical line constructed through the top left vertex of the rotated square.

These three lines will form a larger square that will contain four congruent triangles A, B, C, and D.
Each of these triangles will have sides a, b, and c.
A side of the larger square is $(a + b)$.

Therefore, the area of the larger square is $(a + b)^2$.

The area of each of the four congruent right triangles is $\dfrac{ab}{2}$.

Area of rotated square = area of larger square – area of the four triangles A, B, C, and D.

$$c^2 = (a + b)^2 - 4\left(\dfrac{ab}{2}\right)$$
$$c^2 = a^2 + 2ab + b^2 - 2ab$$
$$c^2 = a^2 + b^2$$

Proof 13

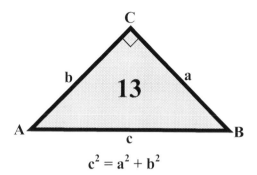

$$c^2 = a^2 + b^2$$

15.13
Isosceles Trapezoid

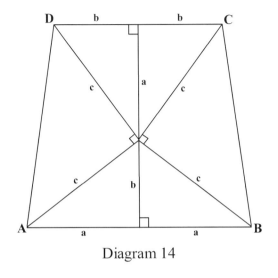

Diagram 14

Reference Diagram 14.

The **plan** is to determine the area of the isosceles trapezoid by two different methods, and then equate the two areas to verify the Pythagorean relationship.

Given:
Isosceles trapezoid ABCD with side AD equal to side BC.
Trapezoid ABCD is comprised of four congruent right triangles with sides a, b, and c, and two congruent isosceles right triangles with sides c, c, AD, and c, c, BC.

Prove:
$$c^2 = a^2 + b^2$$

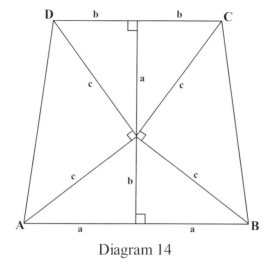

Diagram 14

Reference Diagram 14.

The Area of trapezoid ABCD by the trapezoid formula:

$$A_{ABCD} = \left(\frac{h}{2}\right)(Base\,1 + Base\,2)$$

$$A_{ABCD} = \left(\frac{a+b}{2}\right)(2a + 2b)$$

$$A_{ABCD} = \left(\frac{2a^2 + 2ab + 2ab + 2b^2}{2}\right)$$

(1) $A_{ABCD} = a^2 + 2ab + b^2$

The Area of the six triangles that comprise the trapezoid:

$$4\left(\frac{ab}{2}\right) + 2\left(\frac{c^2}{2}\right)$$

(2) $2ab + c^2$

Equating (1) and (2):
$$a^2 + 2ab + b^2 = 2ab + c^2$$
$$c^2 = a^2 + b^2$$

73

Proof 14

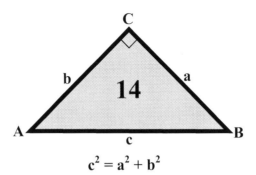

$$c^2 = a^2 + b^2$$

15.14
Inscribed Circle

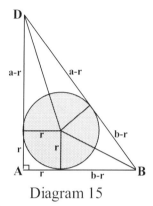

Diagram 15

Reference Diagram 15.

The **plan** is to determine the area of triangle ABD in two different ways, and then equate the two areas to verify the Pythagorean relationship.

Given:
Triangle ABD with right $\angle A = 90°$.
A circle with radius r is inscribed in triangle ABD.
Triangle ABD is comprised of two different pairs of congruent right triangles, and a square with side r that is also the radius of the circle.

Prove:
$$c^2 = a^2 + b^2$$

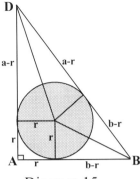

Diagram 15

Reference Diagram 15.

Note that in triangle ABD:

Base, $AB = r + (b - r) = b$,

Height, $AD = r + (a - r) = a$, and the

Hypotenuse, $BD = (a - r) + (b - r) = a + b - 2r = c$

Determine the area of triangle ABD by using the area of a triangle formula:

$$A_{ABD} = \frac{(AB)(AD)}{2}$$

$$A_{ABD} = \frac{(r + (b - r))(r + (a - r))}{2}$$

$$A_{ABD} = \frac{(b)(a)}{2}$$

(1) $A_{ABD} = \dfrac{ab}{2}$

Determine the area of triangle ABD by finding the total area of the five polygons comprising triangle ABD:

$$A_{ABD} = r(a - r) + r(b - r) + r^2$$

$$A_{ABD} = ar + br - r^2$$

(2) $A_{ABD} = r(a + b - r)$

It is necessary to involve c in the analysis:

$$c = (a - r) + (b - r)$$

$$c = a + b - 2r$$

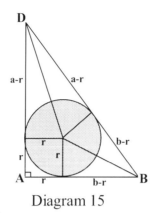

Diagram 15

Reference Diagram 15.

Determine the semiperimeter, p, of triangle ABD:

$$p = \frac{a+b+(a+b-2r)}{2}$$

$$p = a+b-r$$

Note that p is multiplied by r in (2) above, so that

(3) $A_{ABD} = rp$

$$p-c = (a+b-r)-(a+b-2r)$$
$$p-c = a+b-r-a-b+2r$$
$$p-c = r$$

Substituting $(p-c)$ for r in (3),

(4) $A_{ABD} = (p-c)(p)$

Equating (1) and (4):

$$(p-c)(p) = \frac{ab}{2}$$

$$\left(\frac{a+b+c}{2}-c\right)\left(\frac{a+b+c}{2}\right) = \frac{ab}{2}$$

$$\left(\frac{a+b+c-2c}{2}\right)\left(\frac{a+b+c}{2}\right) = \frac{ab}{2}$$

$$(a+b-c)\left(\frac{a+b+c}{2}\right) = ab$$

$$(a+b-c)(a+b+c) = 2ab$$

$$a^2 + 2ab + b^2 - c^2 = 2ab$$

$$c^2 = a^2 + b^2$$

Proof 15

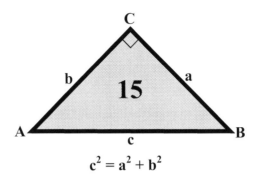

$$c^2 = a^2 + b^2$$

15.15
Merlen Hexagon

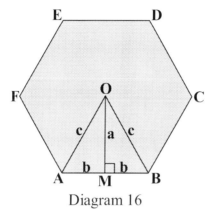

Diagram 16

Reference Diagram 16.

The **plan** is to determine the area of hexagon ABCDEF by two different methods, and then equate the two areas to validate the Pythagorean Theorem.

Given:
Regular hexagon ABCDEF with center O.
Construct triangle AOB with radii OA and OB.
In triangle AOB, let OA and OB equal c.
Construct a perpendicular OM to AB.
Let OM equal a, the altitude of triangle AOB.
The perpendicular a also bisects AB into two equal segments, AM and MB. Let these two equal segments equal b.

Note that there are a total of six of these equilateral triangles that comprise the regular hexagon. Each of these triangles has a central angle of $60°$, and because of the two equal radii, the base angles are also $60°$. Because all three angles of each of the six triangles are equal, each of the six triangles are equilateral.

Prove:
$$c^2 = a^2 + b^2$$

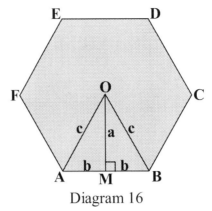

Diagram 16

Reference Diagram 16.

Determine the area of the regular hexagon ABCDEF by applying the formula, $A = \dfrac{ap}{2}$, where a is the apothem of the polygon, and p is the perimeter of the polygon.

Specifically, in Diagram 16:

The apothem is a, and the perimeter of the hexagon is:

$$(6)(2b) = 12b$$

$$A_{ABCDEF} = \frac{(a)(12b)}{2}$$

(1) $A_{ABCDEF} = 6ab$

Determine the area of the triangle AOB by applying Heron's formula:

$$A = \sqrt{(s)(s-a)(s-b)(s-c)}$$

In this formula, s represents half the perimeter (semiperimeter) of a triangle, and a, b, and c represent the three sides of a triangle.

$$s = \frac{a+b+c}{2}$$

In Diagram 16,

$$s = \frac{2b + 2c}{2}$$

$$s = b + c$$

Therefore,

$$A_{AOB} = \sqrt{(b+c)(b+c-2b)(b+c-c)(b+c-c)}$$

$$A_{AOB} = \sqrt{(b+c)(c-b)(b)(b)}$$

$$A_{AOB} = \sqrt{(b+c)(c-b)(b^2)}$$

$$A_{AOB} = b\sqrt{bc-b^2+c^2-bc}$$

$$A_{AOB} = b\sqrt{c^2-b^2}$$

Area of hexagon:

$$A_{ABCDEF} = (6)(A_{AOB})$$

(2) $6A_{AOB} = 6b\sqrt{c^2-b^2}$

Equating (1) with (2):

$$6b\sqrt{c^2-b^2} = 6ab$$

$$(36b^2)(c^2-b^2) = 36a^2b^2$$

$$36b^2c^2 - 36b^4 = 36a^2b^2$$

$$b^2c^2 - b^4 = a^2b^2$$

$$b^2c^2 = a^2b^2 + b^4$$

$$c^2 = a^2 + b^2$$

Proof 15.15v

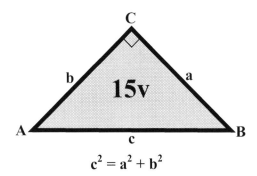

$$c^2 = a^2 + b^2$$

15.15v
Square of Merlen

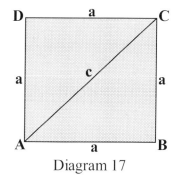

Diagram 17

Reference Diagram 17.

Note that 15.15v is a variation of 15.15 because the analysis is similar, but a quadrilateral is involved instead of a hexagon.

The **plan** is to determine the area of square ABCD by two different methods, and then equate the two areas to validate the Pythagorean Theorem.

Given:
Square ABCD with side a.
Construct diagonal AC.
Let AC equal c.

Prove:
$$c^2 = a^2 + b^2$$

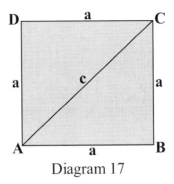
Diagram 17

Reference Diagram 17.

Determine the area of square ABCD by the area formula:

(1) $A_{ABCD} = a^2$

Determine the area of triangle ABC by applying Heron's formula:

$$A = \sqrt{(s)(s-a)(s-b)(s-c)}$$

In this formula, s represents half the perimeter (semiperimeter) of a triangle, and a, b, and c represent the three sides of a triangle.

$$s = \frac{a+b+c}{2}$$

$$s = \frac{2a+c}{2}$$

Therefore,

$$A_{ABC} = \sqrt{\left(\frac{2a+c}{2}\right)\left(\frac{2a+c}{2}-a\right)\left(\frac{2a+c}{2}-a\right)\left(\frac{2a+c}{2}-c\right)}$$

$$A_{ABC} = \sqrt{\left(\frac{2a+c}{2}\right)\left(\frac{c}{2}\right)\left(\frac{c}{2}\right)\left(\frac{2a-c}{2}\right)}$$

$$A_{ABC} = \sqrt{\left(\frac{c^2}{4}\right)\left(\frac{2a+c}{2}\right)\left(\frac{2a-c}{2}\right)}$$

$$A_{ABC} = \frac{c}{2}\sqrt{\frac{4a^2-c^2}{4}}$$

$$A_{ABC} = \frac{c}{4}\sqrt{4a^2-c^2}$$

Square ABCD is comprised of two congruent triangles ABC and ADC.

$$A_{ABCD} = 2\left(\frac{c}{4}\sqrt{4a^2 - c^2}\right)$$

(2) $A_{ABCD} = \dfrac{c}{2}\sqrt{4a^2 - c^2}$

Equating (1) and (2):

$$\frac{c}{2}\sqrt{4a^2 - c^2} = a^2$$

$$\frac{c^2}{4}\left(4a^2 - c^2\right) = a^4$$

$$c^2\left(4a^2 - c^2\right) = 4a^4$$

$$4a^2c^2 - c^4 = 4a^4$$

$$c^4 - 4a^2c^2 + 4a^4 = 0$$

$$\left(c^2 - 2a^2\right)\left(c^2 - 2a^2\right) = 0$$

$$c^2 - 2a^2 = 0$$

$$c^2 = 2a^2$$

$$c^2 = a^2 + a^2$$

Because two sides of the triangle are equal, that is, $a = b$,

$$c^2 = a^2 + b^2$$

16
Peculiar Pythagorean Perspective

There were three Indian squaws. One squaw slept on a deer skin. One slept on an elk skin, and the third slept on a Hippopotamus skin. All three became pregnant, and the first two each had a baby boy. The squaw who slept on the hippopotamus skin had twin boys. This proves that **the squaw of the hippopotamus is equal to the sons of the squaws of the other two hides.**

17
Puzzle

Apparently, 64 square inches can be transformed into 65 square inches by cutting a figure into four pieces, and assembling them in a different arrangement. Do you think such a thing is possible?

Using graph paper, cut out a square 8 units on a side. Use a unit large enough to work with easily. Then cut along the lines as indicated in Diagram 1, resulting in two triangles and two quadrilaterals. Rearrange the pieces as shown in Diagram 2. This new diagram appears to be a rectangle with length of 13 units, and a width of 5 units. Its area is 65 square units compared to the original 64 square units of Diagram 1.
Is something wrong? If so, what is it?

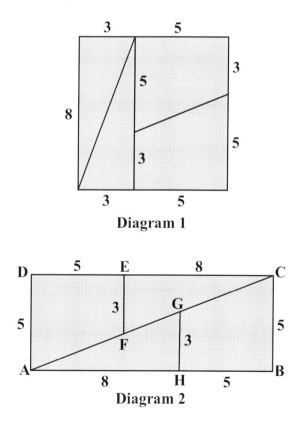

Diagram 1

Diagram 2

Hint: How many degrees are there in ∠ECF? in ∠GCB?

18
Questions & Answers

Question
1
Triangle Median

The median from any vertex of any triangle forms two other triangles. What is the relationship of these two triangles?
Can you provide proof?

Consider the diagram below:

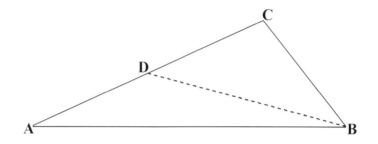

Given:
Triangle ABC.
Median BD bisects AC so that AD equals DC.
BD forms triangles ABD and BDC.

Question:
What is the relationship between triangles ABD and BDC?

[Reference the Appendix for the answer].

Question
2
Hunter & Bear

A hunter left his campsite, walked directly south exactly 1 mile. He then turned and walked directly east exactly 3 miles. At that location, he shot a bear. Then, after walking north exactly 1 mile, the hunter was back at his campsite.

Question:
What **color** was the bear?

[Reference the Appendix for the answer].

Question
3
Right Triangle Median

In a right triangle, the median from the vertex of the right angle is one-half of the hypotenuse.
Can you prove this statement?

Consider the diagram below:

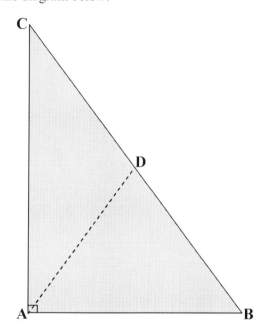

Given:
Triangle ABC.
Median AD from vertex A to BC bisects BC so that BD equals CD.

Question:
Can you prove that AD is half of BC?

[Reference the Appendix for the answer].

Question
4
Four States

There are four states in our contiguous forty-eight United States whose boundaries meet at a point forming four <u>right angles</u>. No other configuration of states can claim such a geographical distinction.

Question:
Can you identify the four states?

[Reference the Appendix for the answer].

Question 5 Challenge

Consider the Diagram below:

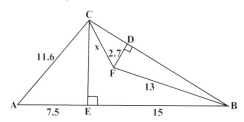

Given:
In the multiple triangle configuration,
AE = 7.5
BE = 15
AC = 11.6
BF = 13
DF = 2.7
CF = x
$\angle BEC$ and $\angle BDF = 90°$

Question:
Determine the value of x to the nearest tenth.

[Reference the Appendix for the answer].

Question
6
Pleasant Triangles

Question:
Why is a Right Triangle more pleasing than any other type of triangle?

[Reference the Appendix for the answer].

Question
7
Finding Something

Consider the details of the diagram illustrated below:

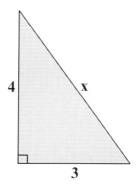

Question:
In the diagram above, can you find x?

[Reference the Appendix for the answer].

Question
8
Moral Angles

Question:
What type of triangle demonstrates good moral character?

[Reference the Appendix for the answer].

Question
9
Volume of a Sphere

This is an application of the Pythagorean Theorem involving three dimensions as presented in Chapter 12.

Given:
A sphere is inscribed in a cube.
This also means that the cube is
circumscribed about the sphere.
The diagonal (AB) of the cube

face is $3\sqrt{6}$ meters.

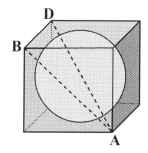

What is the volume of the sphere (nearest tenth of a meter)?

[Reference the Appendix for the answer].

Epilogue

I want to share some information regarding two significant aspects of my life:

(A) My Right-Angled Friends
(B) My Dual Citizenship

(A) My Right-Angled Friends

On my dedication page of this book, I mentioned my "right-angled" friends. These friends are not "right" because of their intelligence, personality, success, comeliness, or other favorable characteristics. These friends are <u>right</u> because they have demonstrated the correct, specific perspective by which they acknowledge and perceive Reality. This Reality is the Lord Jesus, the Creator, Sustainer, and Redeemer of <u>all</u> Reality. Jesus is the essence of all that <u>is</u> Right, and He is the Savior and Lord of all who will trust Him. Is He your Savior? If not, you must make Him the first priority in your life. Your eternal future depends upon your response to Jesus. He is God's <u>only</u> provision for you to become <u>right</u>.

(B) My Dual Citizenship

(1) I am a citizen of the United States.
   This is my temporary home.
   I am just a pilgrim passing through this world.
   As an Ambassador for Jesus, my mission is to represent Him.

(2) I am a Citizen of God's Heavenly Kingdom.
   I am a Permanent Resident with Jesus.
   I will have Fellowship Forever with my Savior, the Lord Jesus.
   Heaven is my Eternal Future.

Appendix

Answers to Questions of Chapter 18

Question
1
Triangle Median

The median from any vertex of any triangle forms two other triangles. What is the relationship of these two triangles? Can you provide proof?

Consider the diagram below:

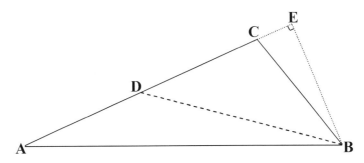

Given:
Triangle ABC.
Median BD bisects AC so that AD equals DC.
BD forms triangles ABD and BDC.

Question:
What is the relationship between triangles ABD and BDC?

QAQAQAQAQAQAQAQAQAQAQAQAQAQAQAQAQA

Answer:
Extend AC.
Construct a perpendicular BE from B to AC extended.
BE is the same altitude for both triangle ABD and triangle BDC.
Triangles ABD and BDC have equal bases, AD and DC respectively.
Because triangles ABD and BDC have equal bases and altitudes, they also have <u>equal</u> areas.

Question 2
Hunter & Bear

A hunter left his campsite, walked directly south exactly 1 mile. He then turned and walked directly east exactly 3 miles. At that location, he shot a bear. Then, after walking north exactly 1 mile, the hunter was back at his campsite.

Question:
What color was the bear?

QAQAQAQAQAQAQAQAQAQAQAQAQAQAQAQA

Answer:
An exact North Pole location is the only location where you can travel any distance southward (the only direction you can travel), then any distance eastward or westward, and then the same distance northward as you traveled southward, and return to your starting location.
The bear had to be a polar bear.
Obviously, the color of the bear was white. ☺

Question
3
Right Triangle Median

In a right triangle, the median from the vertex of the right angle is one-half of the hypotenuse.
Can you prove this statement?

Consider the diagram below:

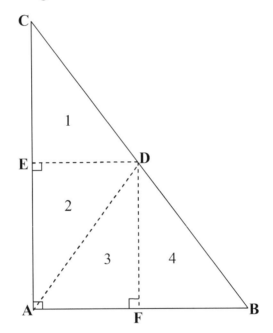

Given:
Triangle ABC.
Median AD from vertex A to BC bisects BC so that BD equals CD.

Question:
Can you prove that AD is half of BC?

QAQAQAQAQAQAQAQAQAQAQAQAQAQAQAQAQA

Answer:

Why is D equally distant from all three vertices (A,B,C) of triangle ABC? The answer to this question is not intuitively obvious.

The three <u>medians</u> of any triangle are concurrent, and intersect at a point inside a triangle, but this point of concurrency is not necessarily equally distant from all three vertices.

The three <u>bisectors</u> of the angles of a triangle are also concurrent, and intersect at a point that is equally distant from the three sides of the triangle. This point is known as the <u>incenter</u>, and is the center of a circle that can be inscribed in the triangle.

The <u>perpendicular bisectors</u> of the three sides of a triangle are also concurrent, and intersect at a point that is <u>equally distant from the three vertices of a triangle</u>.

Therefore, point D must be the point of concurrency where the three perpendicular bisectors meet. If the triangle is <u>acute</u>, the perpendicular bisectors meet at a point inside the triangle. If the triangle is <u>obtuse</u>, the perpendicular bisectors meet at a point outside the triangle. If the triangle is a <u>right</u> triangle, the perpendicular bisectors meet at the midpoint of the triangle's hypotenuse.

Construct line DE perpendicular to AC.
Construct line DF perpendicular to AB.
These two constructions form congruent right triangles 1, 2, 3, and 4.

Triangles 2 and 3 share a common hypotenuse, so there are three equal hypotenuses. Corresponding sides and angles of congruent polygons are equal. In this specific case, all three hypotenuses of the four triangles are equal, and point D is equally distant from the three vertices of triangle ABC. In other words, D is the center of a circle that will intersect vertices A, B, and C. The circle is referred to as the circumcircle of triangle ABC.

Conclusion:
The midpoint of the hypotenuse of a right triangle is equally distant from all three vertices of the triangle.

Question
4
Four States

There are four states in our contiguous forty-eight United States whose boundaries meet at a point forming four <u>right angles</u>. No other configuration of states can claim such a geographical distinction.

Question:
Can you identify the four states?

QAQAQAQAQAQAQAQAQAQAQAQAQAQAQAQA

Answer:

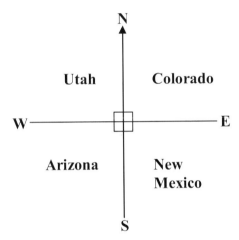

Question 5 Challenge

Consider the Diagram below:

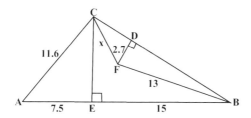

Given:
In the multiple triangle configuration,
AE = 7.5
BE = 15
AC = 11.6
BF = 13
DF = 2.7
CF = x
$\angle BEC$ and $\angle BDF = 90°$

Question:
Determine the value of x to the nearest tenth.

QAQAQAQAQAQAQAQAQAQAQAQAQAQAQAQA

Answer

$$\left(CE\right)^2 = \left(AC\right)^2 - \left(AE\right)^2$$

$$CE = \sqrt{\left(AC\right)^2 - \left(AE\right)^2}$$

$$CE = \sqrt{\left(11.6\right)^2 - \left(7.5\right)^2}$$

$$CE = 8.8$$

$$\left(BC\right)^2 = \left(BE\right)^2 + \left(CE\right)^2$$

$$BC = \sqrt{\left(BE\right)^2 + \left(CE\right)^2}$$

$$BC = \sqrt{\left(15\right)^2 + \left(8.8\right)^2}$$

$$BC = 17.4$$

$$(BD)^2 = (BF)^2 - (DF)^2$$

$$BD = \sqrt{(BF)^2 - (DF)^2}$$

$$BD = \sqrt{(13)^2 - (2.7)^2}$$

$$BD = 12.7$$

$$CD = BC - BD$$

$$CD = 17.4 - 12.7$$

$$CD = 4.7$$

$$x^2 = (DF)^2 + (CD)^2$$

$$x = \sqrt{(DF)^2 + (CD)^2}$$

$$x = \sqrt{(2.7)^2 + (4.7)^2}$$

$$x = 5.4$$

Question
6
Pleasant Triangles

Question:
Why is a Right Triangle more pleasing than any other type of triangle?

QAQAQAQAQAQAQAQAQAQAQAQAQAQAQAQA

Answer:
A Right Triangle is more pleasing than other types of triangles, because
A Right Triangle has two complementary angles! ☺

Question
7
Finding Something

Consider the details of the diagram illustrated below:

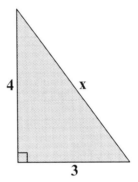

Question:
In the diagram above, can you find x?

QAQAQAQAQAQAQAQAQAQAQAQAQAQAQAQA

Answer:

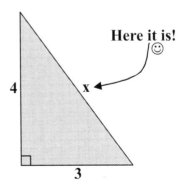

Postscript:
If your sensibility has been offended,
The value of x is 5.

Question
8
Moral Angles

Question:
What type of triangle demonstrates good moral character?

QAQAQAQAQAQAQAQAQAQAQAQAQAQAQAQA

Answer:
A <u>Right</u> Triangle is naturally superior to a <u>wrong</u> triangle.☺

Question 9
Volume of a Sphere

Given:
A sphere is inscribed in a cube.
This also means that the cube is
circumscribed about the sphere.
The diagonal (AB) of the cube
face is $3\sqrt{6}$ meters.

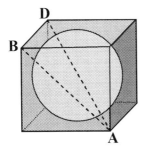

What is the volume of the sphere (nearest tenth of a meter)?

QAQAQAQAQAQAQAQAQAQAQAQAQAQAQAQAQA

Answer:
Let e be an edge of the cube.
Applying the Pythagorean Theorem to a face of the cube:

$$\left(3\sqrt{6}\right)^2 = e^2 + e^2$$
$$54 = 2e^2$$
$$27 = e^2$$
$$\sqrt{27} = e$$
$$3\sqrt{3} = e$$

Now e is also the diameter of the sphere.

$$\frac{e}{2} = \text{radius of the sphere.}$$

$$\frac{3\sqrt{3}}{2} = \text{radius of the sphere}$$

Volume of sphere:

$$V = \frac{4}{3}\pi r^3$$

$$V = \frac{4}{3}\pi \left(\frac{3\sqrt{3}}{2}\right)^3$$

$$V = \frac{4}{3}\pi \left(\frac{81\sqrt{3}}{8}\right)$$

$$V = 4\pi \left(\frac{27\sqrt{3}}{8}\right)$$

$$V = \pi \left(\frac{27\sqrt{3}}{2}\right)$$

$$V = 73.5 \; meters^3$$

Printed in the United States
By Bookmasters